PRACTICAL DJANGO

learn with examples

Upendra Handa

Copyright © 2021 WonderBlunders

All rights reserved

The characters and events portrayed in this book are fictitious. Any similarity to real persons, living or dead, is coincidental and not intended by the author.

No part of this book may be reproduced, or stored in a retrieval system, or transmitted in any form or by any means, electronic, mechanical, photocopying, recording, or otherwise, without express written permission of the publisher.

ISBN: 9798520197133

Cover design by: Art Painter
Library of Congress Control Number: 2018675309
Printed in the United States of America

dedicated to my family and everyone who inspired me

"Imagintion is more important than knowledge"

- Albert Einstein

So, whatever you learn, you should be able to imagine and predict the scenario if you have really understood it.

CONTENTS

Title Page
Copyright
Dedication
Epigraph
Foreword
Introduction
Request Response Cycle — 1
Practice 101 — 15
Static Files and Templates — 23
Practice 201 — 39

FOREWORD

This book is for everyone who wants to learn Django from scratch. I have tried to put forward many examples and analogies to relate concepts with what we already know.

I request readers to first grasp the concepts and not to worry about the syntax too much. Whenever we learn something new, we get anxious about the syntax and miss the concepts because of it. Try to just know the vague meaning of the syntax and not to memorize every syntax as it comes. Practice will help you later to do that automatically.

INTRODUCTION

In the first few chapters, we will learn about basics of how internet and website work together. You will familarize yourself with terms like client, browser, request, url, dns, port number, server, response etc.

In later chapters, we dive deep into how Django specifically works to generate response that is then served by server back to the client.

Overall, you will learn how to build websites and/or json content using MVT and standard web technologies like HTML, CSS, jQuery and Ajax.

REQUEST RESPONSE CYCLE

*an overview of the journey of
a client request to the server
and response from the server*

Introduction

When a user types a web address in the web browser like www.example.com/index.html, he is essentially trying to find a resource (index.html in this case) on the cloud with an address. It is more or less similar to how a delivery person tries to locate a person at a given address.

Since it is hard for humans to remember numbers but easier to remember names, every machine on the cloud, which is acutally identified by an IP address, for example 4.4.4.4, is given a name like www.google.com. Whenever a client requests a website name, its corresponding IP address is resolved first and then the correct machine is traced. The resource is requested from this machine (server).

Whatever is returned by the server in reply to a request is called a response. This response can be a file like a pdf file, an html page or some plain-text like json, xml etc.

In simple words, all what above text means is that a client types in an address of a resource he is requesting, and the address is used to find the server which sends back a response in reply to his request. This response can vary from an html page, to a file, to plain-text etc.

Url Patterns

Imagine www.example.com is just a folder on the cloud. Similarly, www.google.com is another folder on the cloud. We will call it root folder because it contains all the sub-folders and files for that website, so it is the main folder.

Now, the address (url) may look like this:

www.example.com/index.html

or this:

www.example.com/index

In the first one, the user is writing down the name of the file (index.html) and location (root folder). The resource can be located easily here (an index.html file inside root folder).

root folder -> index.html

Another example - www.example.com/temp/a.html. Here, the resource a.html file is located inside a folder temp which is again inside the root folder. In other words:

root folder -> temp -> a.html

The second example, *www.example.com/index* is actually referring to a pattern and not the direct path of the file.

URL -> www.example.com/index
URL Pattern -> index

A url pattern does not include www.example.com. Another example of url pattern is:

URL -> www.example.com/files/abc.
URL Pattern -> files/abc

Django uses patterns to locate resources. Django contains a file called urls.py (structure of Django project will be discussed in detail soon). This file contains patterns matched to functions. For example, *index* is a pattern, so urls.py will just tell which function to run when the pattern requested is *index*. This function will, in turn, return the response, which can be an html file, or a pdf file, or a plain-text etc.

Comparing these two ways of finding resources on the internet, we see that the one with pattern does not reveal the name of the resource or its location. And Django uses pattern-based approach for requests.

To summarize, in Django:

url pattern -> server -> urls.py -> function -> response

Practical Learning:

While programming in Django, the very first few things we do before writing function code is to write an entry in urls.py.

For example, say you want to make a website page called homepage. It is stored as homepage.html. All resources in Django are returned via functions. In simple words, we will write a function called homepage, which will return homepage.html.

Now, as we discussed before, job of urls.py is to tell which function has to run on which url pattern received from user request url. It is totally upto us to decide the pattern which will return homepage function. Let's say, we decided that the url we want to serve homepage to is www.example.com/homepage. So, the pattern inside this url is *homepage* (a url pattern is not the whole url but everything except the www.example.com).

The entry inside urls.py will look like this:

url(r'^homepage', views.homepage, name='homepage')

It means, a url pattern which looks like *homepage*, will run a function homepage located inside views file (views.homepage). Forget the third part, *name*, for now.

Returning First Page Via Django

Since we now know that Django can identify a url pattern from a user request and return a resource (say, an html file), why not try it rightaway?

For this, following flow will be followed:

STEP 1: An OS (Ubuntu in my case) will be required to install Python (version 3).

STEP 2: Once Python is installed, we will install a virtual environment. The meaning and need of virtual environment is explained in next paragraph.

STEP 3: Inside a virtual environment, we will install the latest Django framework using package manager pip.

STEP 4: That's it. We have Django installed on our system. All we need to do now is to create an empty project in Django, then an app inside that project and configure a few things (like writing an entry in urls.py, writing a function for homepage etc).

What is Virtual Environment?

Using Django, we are going to make a lot of projects in future. Each project will require a different set of libraries to be installed to help in our programming. At the end of each project, we make a file called requirements.txt which contains the name of all the libraries used by our project. This is to help us transfer our project somewhere else, so we can quickly install all those libraries and make our project working on a different machine too.

Now imagine, you have two projects A and B.
A requires libraries 1, 2, 3, 4 and 5.
B requires libraries 4, 5, 6, 7 and 8.
As we can see, both A and B require libraries 4 and 5 in common. But rest of them are different. So if we install all the libraries required by project A in our operating system, we will have 1, 2, 3, 4 and 5 installed in our OS. Now, if we also install the libraries required by project B, we already have 4 and 5 installed. So, our system will install 6, 7 and 8 additionally.

Now comes the part to ponder about. When we will make requirements.txt file for project A, it will write down libraries 1, 2, 3, 4, 5, 6, 7, and 8. It simply checks the libraries installed and write them down. We can observe that 6, 7 and 8 are not required by project A. Similarly, requirements.txt file of project B will have the same libraries 1, 2, 3, 4, 5, 6, 7 and 8 and project B does not require libraries 1, 2 and 3 either.

To solve above problem, each project is run inside a virtual environment, and every library installed inside that virtual environment remains inside it only. So, each such virtual environment of a project is a separate package for that project only with only the libraries installed for that project alone. Now, when we make requirements.txt file from inside a virtual environment, it will register only libraries installed inside that environment and not the whole OS. Think of virtual environment as a separate folder that contains your project as well as libraries for that project.

STEP 1 - Install Python

You can check version of Python installed on your Ubuntu OS with the following command:

python --version

If not installed, you can find any good tutorial online to install Python 3 on your Ubuntu OS. For example - *https://phoenixnap.com/kb/how-to-install-python-3-ubuntu*

On Windows OS, Python 3 can be installed by downloading the exe installer from *https://www.python.org/downloads/*

STEP 2 - Install Virtual Environment

On Ubuntu terminal, we type the following command to install virtual environment:

sudo apt-get install python3-venv

Now, let's create a virtual environment named *env* in our current directory:

python3 -m venv env

Once we have successfully created the virtual environment, we will activate it so that all activities we do like installing libraries etc will be limited to that virtual environment only:

source env/bin/activate

Great! Now we can go ahead and install Django in this virtual environment named *env*.

STEP 3 - Install Django inside Virtual Environment

Issue the following command:

pip3 install django

STEP 4 - Configuration and Coding

Here's the exciting part now. As discussed earlier, we first need to create an empty Django project to work on.

django-admin startproject myfirstproject

A project is a collection of various apps installed inside it. This division is done to handle different tasks of our project inside different apps. So, let's make one such app inside our project.

First, we go inside the project directory:

cd myfirstproject

Now, we create an app inside it:

python manage.py startapp myfirstapp

We need to register each app with our project in order to enable it. So, we go inside the settings.py file inside the project folder and add the following:

INSTALLED_APPS = [
  'django.contrib.admin',
  'django.contrib.auth',
  'django.contrib.contenttypes',
  'django.contrib.sessions',
  'django.contrib.messages',
  'django.contrib.staticfiles',
  ...
  # Add our app here

    'myfirstapp',
]

The text written above after # is called comment. It is ignored and is not treated as code.

We have registered our app myfirstapp. One last thing we need to do is related to location of our html files. Since our aim is to return an html file as a response, our application must know the default folder where it should look for all those html files to return because we cannot keep writing the full path everytime we want to return an html file.

Django offers a simple solution for it. In settings.py file, we have this:

```
TEMPLATES = [
  {
    'BACKEND': 'django.template.backends.django.DjangoTemplates',
    'DIRS': [BASE_DIR / 'templates'],
    'APP_DIRS': True,
    ...
  },
]
```

Above setting tells our project where to look for html template files.

Focus on last two lines: DIRS and APP_DIRS. These two are the default folder locations for html file templates.

DIRS refer to the folder where we want to keep the common html files (just for conceptual clarification). Say, you want to show a homepage, so you can keep the homepage html file in this folder.

APP_DIRS, if set to value True, means that our project will look for html files in a folder named *templates* inside each app we create in our project. This helps us to serve html files of each app separately from that *templates* folder.

For example, say you want to serve an html file named *app-page.html* from myfirstapp. You will then have to keep this file like this:

myfirstapp -> templates -> myfirstapp -> apppage.html

So, whenever you will request *myfistapp/apppage.html,* it will be served from this path.

Why the same name folder inside templates folder?

You might be curious, why do we have to create another folder named *myfirsapp* inside *templates* folder and why not simply put apppage.html inside *templates* folder?

Imagine you have two apps, *myfirstapp* and *mysecondapp.* You have *apppage.html* file in both of them. How will you be able to distinguish between them? That is why, we create a folder inside *templates* folder (you can choose a custom name for this folder but as a standard, we name it same as the app name). And whenever we want to return *apppage.html* file, we also mention from which particular folder to return it, like:

return myfirstapp/apppage.html

return mysecondapp/apppage.html

To summarize:

We have created a virtual environment with Django installed inside it. We have also created a Django project and an app inside that project. After registering that app inside settings.py and having a look at Templates settings, we now know that we can keep our app-specific html files in the following path:

myfirstapp -> templates -> myfirstapp -> apppage.html

We can also keep common html files inside a *templates* folder located in the root project folder.

So What is Going On?

Before we move further with coding, we should take a break and observe how Django has planned its directory structure.

First, we learned that there are two locations for html files - one is app-specific templates folder inside each app and second is a common folder named templates inside project root folder.

If you go to your project root folder, you will notice a folder named myfirstproject. It is the default app created whenever we create a new project, with the same name as our project has. It is already registered by default.

So why does Django create this default app and why does it keep two different locations for html files. Well, to put it simply, we will just do this. We will do all common tasks inside default app only and any html file belonging to that task will be kept inside the common templates folder. This is for conceptual clarification that each app has its own template folder, even our default app too! One example of a common task is to show a landing page or a homepage (which is the task we are after too).

CODING

Going back to our aim, we wanted to serve an html file named homepage. Since, it is not specific to any app, we will put this file inside the common html folder *templates* inside the project root directory. Let's write 'Hello World' inside that html file and save it in the templates folder.

myfirstproject -> templates -> homepage.html

From our previous discussion, we know that we need to write a function to return this html file and then write a corresponding entry in urls.py for the function. So, let's go to the urls.py of default app myfirstproject.

from . import views

urlpatterns = [
url(r'^homepage', views.homepage, name='homepage'),
]

Let's create a file called views.py inside the default app and write the function named homepage to return the html file inside it.

def homepage(request):
    return render(request, 'myfirstproject/homepage.html', {})

That's it! Now go to the terminal where we were writing commands and type:

python manage.py runserver

and the Django will return your html file when you point your browser to 127.0.0.1:8000/homepage (similar to www.example.com/homepage).

PRACTICE 101

more examples to practice request-response cycle

Return an html file from default app

(previous example)

This is the same example of returning homepage.html file from the default myfirstproject app as discussed in last ending section.

i. Place the homepage.html file in the default template folder (inside project root folder)

ii. Write a function in views.py file that will return this homepage.html file

def homepage(request):
    return render(request, 'myfirstproject/homepage.html', {})

iii. Make an entry in urls.py file to call above function whenever a user url is www.example.com/homepage

Url - www.example.com/homepage
Url Pattern - homepage

urls.py:

from . import views

urlpatterns = [
url(r'^homepage', views.homepage, name='homepage'),
]

Return an html file from myfirstapp

We know that app-specific html folder is named templates which is found inside each app folder (in our case, myfirstapp) and html file is placed like this inside it:

myfirstapp -> templates -> myfirstapp -> apppage.html

i. Place the apppage.html file as shown above

ii. Write a function in views.py file *found inside myfirstapp folder* that will return this homepage.html file

def myfirstapppage(request):
return render(request, 'myfirstapp/myfirstapppage.html', {})

iii. Make an entry in urls.py file *found inside myfirstapp folder* to call above function whenever a user url is www.example.com/myfirstapppage

Url - www.example.com/myfirstapppage
Url Pattern - myfirstapppage

urls.py (inside myfirstapp folder):

from . import views

urlpatterns = [
url(r'^myfirstapppage', views.myfirstapppage, name='myfirstapp-page'),
]

Including Urls In Main Urls.py File

Now, this coding is fine but it will not work. It will only work if the user request will hit urls.py file of myfirstapp. But in practice, user request hits only the urls.py file of the default app, which is myfirstproject and not myfirstapp.

So, we will need to include the urls.py patterns of myfirstapp into the urls.py file of myfirstproject. This is how we do it:

url(r'^', include(('myfirstapp.urls', 'myfirstapp'), namespace='myfirstapp')),

We have to be careful here because the patterns might clash. Say, you have a following pattern match in both myfirstproject and myfirstapp:

url(r'^pdf_download', views.pdf_download, name='pdf_download')

Since myfirstproject will be called first when a user sends a request, so everytime a user sends a request like www.example.com/pdf_download, it will go to pdf_download function of myfirstproject and never to myfirstapp. To solve this, we add a prefix in r'^', like r'^myfirstapp' and the pattern becomes:

url(r'^myfirstapp/', include(('myfirstapp.urls', 'myfirstapp'), namespace='myfirstapp')),

Now, to call the pdf_download of myfirstapp, a user will have to send request with myfirstapp prefix like:

www.example.com/myfirstapp/pdf_download

Returning Plain-text from myfirstapp

Let's say, we just want to return a palin-text *hello* in response to a user request. We will not return any html file in this case but plain-text, with the help of a function called HttpResponse.

i. Write a function in views.py file *found inside myfirstapp folder* that will return *hello*

from django.http import HttpResponse

def myfirsthttpresponse(request):
    return HttpResponse('hello')

ii. Make an entry in urls.py file *found inside myfirstapp folder* to call above function whenever a user url is www.example.com/myfirsthttpresponse

Url - www.example.com/myfirsthttpresponse
Url Pattern - myfirsthttpresponse

urls.py (inside myfirstapp folder):

from . import views

urlpatterns = [
url(r'^myfirsthttpresponse', views.myfirsthttpresponse, name='myfirsthttpresponse'),
]

iii. Include myfirstapp urls in main urls.py found in myfirstproject folder:

url(r'^myfirstapp/', include(('myfirstapp.urls', 'myfirstapp'), namespace='myfirstapp')),

Returning JSON from myfirstapp

JSON is a collection of name-value pairs. Let's say, we want to return JSON in response to a user request. We will not return any html file in this case but JSON string, with the help of a function called JsonResponse.

i. Write a function in views.py file *found inside myfirstapp folder* that will return JSON string:

from django.http import JsonResponse

def myfirstjsonresponse(request):
    return JsonResponse({'name' : 'Binod'})

Note - In Django, we essentially return a dictionary as Json-Response. If we set safe parameter to false, we can return any JSON-serializable object.

ii. Make an entry in urls.py file *found inside myfirstapp folder* to call above function whenever a user url is www.example.com/myfirstjsonresponse

Url - www.example.com/myfirstjsonresponse
Url Pattern - myfirstjsonresponse

urls.py (inside myfirstapp folder):

from . import views

urlpatterns = [
url(r'^myfirstjsonresponse', views.myfirstjsonresponse, name='myfirstjsonresponse'),
]

iii. Include myfirstapp urls in main urls.py found in myfirstproject folder:

```
url(r'^myfirstapp/', include(('myfirstapp.urls', 'myfirstapp'),
namespace='myfirstapp')),
```

STATIC FILES AND TEMPLATES

looking under the hood

Returning Static Files Like Css, Js, Images Etc

So far, we have discussed about how to configure our project to look for html files, and how to return text and json data. But what about css files, js files, images etc? A page without images and design is just so boring.

Just like we configured settings.py with APP_DIRS: True to tell our project to look for a folder named templates inside each app for html files, Django by default looks for a folder named static inside each app for static files, when we call /static/ or use static keyword in html links.

In other words, we can make a folder named static inside each app and put our static files there. For example, if we put a jpg image file, named a.jpg, inside static folder of myfirstapp, we can refer to this file from html like this:

<img src="/static/a.jpg">

or, like:

{% load static %}

<img src="{% static 'a.jpg' %}">

By using /static/ or static keyword inside {% %}, we are simply triggering our project to look for static files. /static/ is a url pattern, so whenever we use it, Django will know that we are asking it to search for static files.
Once called to search for static files, Django will look for folders

defined inside settings to look for static files, like:

STATICFILES_DIRS = [< LIST OF DIRECTORIES HERE >]

If not found there, Django looks for folder named static inside each app by default for static files.

Similarity Between Template Lookup And Staticfiles Lookup

If we observe closely, Django has a similar pattern of lookup for templates and static files. First, it looks for the templates (or static files) inside the directories defined under DIRS (or STATICFILES_DIRS). If not found there, it will go ahead and lookup in a folder named static (templates, in case of templates) inside each app.

So what is STATIC_URL and STATIC_ROOT in settings.py file?

When we use /static/ in a link inside html, it means we are asking our project to look for static files. This /static/ is the url pattern that Django identifies to search for static files. We can change this pattern from inside settings.py file by changing STATIC_URL. Say, we change it to /staticfile/, so next time we want Django to search for static files, we will have to use /staticfile/ in our links instead of /static/.

STATIC_ROOT is simply a single directory path where all static files will be copied when we run a command like this:

python manage.py collectstatic

During development (Debug=True in settings.py), we do not even need STATIC_ROOT setting. Because Django looks-up for the static files in different folders as described in previous paragraphs. But when our project is live (Debug=False), it will be very efficient if we take-off this additional load of searching for static files from Django.

What we can do instead is to copy all our static files in one directory and then tell each request which has /static/ in it, to look for that folder. This task of lookup is not done by Django anymore but Nginx. All static lookups by Django stop and Django feels free to do other tasks. We will learn more about it later.

In other words, after deployment, STATIC_ROOT folder is useful as it takes away the load from Django for searching for static files.

Django Project Structure

Django project structure looks something like below:

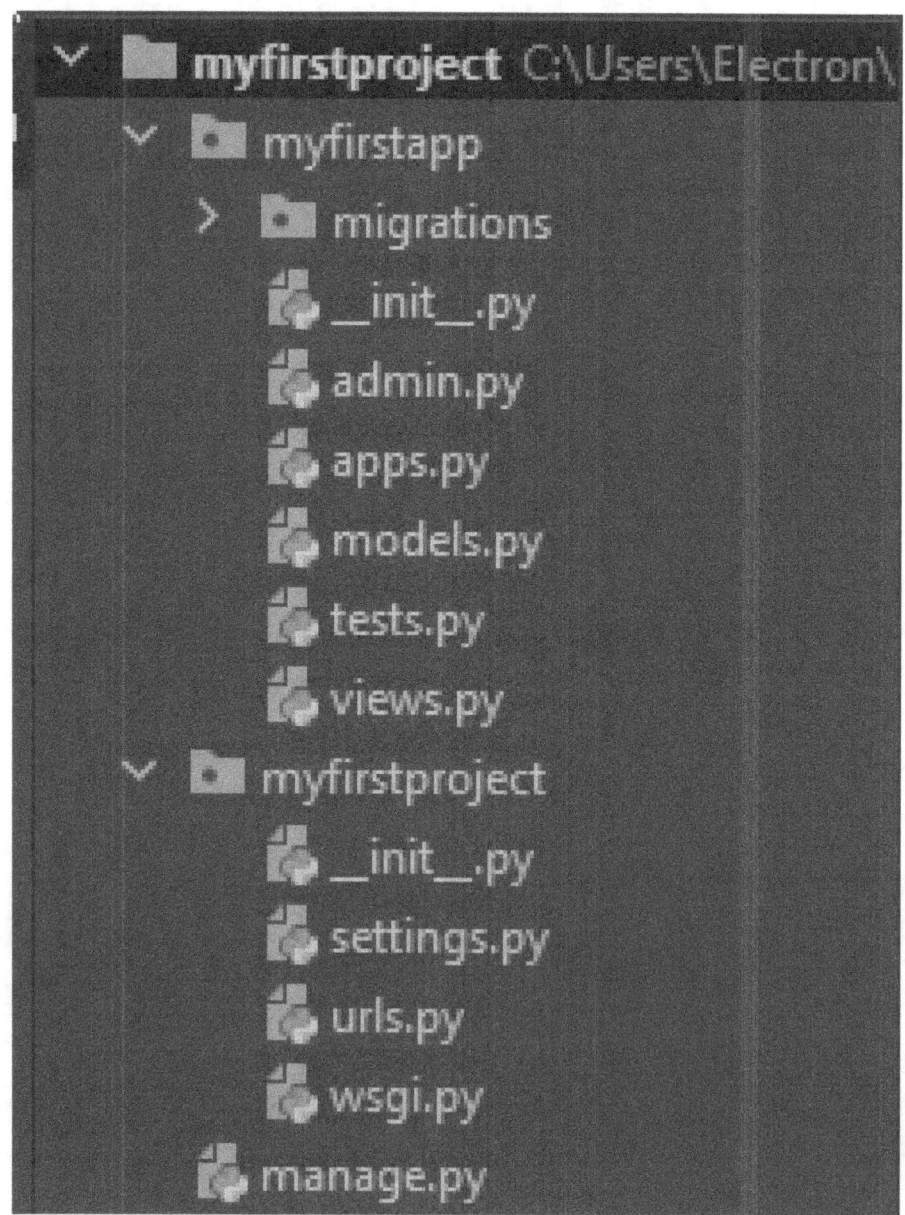

The root project folder named myfirstproject contains a default app, myfirstproject. And then there is another app created by us named myfirstapp.

Each app by default contains some files like models.py, views.py, admin.py etc. In addition, we created a urls.py file inside myfirstapp to contain its url patterns. And then we included it inside main urls.py file of myfirstproject.

We coded a function inside views.py file of myfirstapp to return myfirstapppage.html. We also created a views.py file inside myfirstproject folder and coded a function inside it to return homepage.html.

We can conclude that:

- *views.py file is essentially used to write functions to return responses.*

- *urls.py file is essentially used to map url patterns to functions*

- *models.py file is essentially used to write table structures of database.*

We will now focus on the last part - models.py. We observe that so far, we are just returning static html pages with static files content. To make things more interesting, we need to add database-based content in it. For example, we can collect user messages in database from one page and show them all in a separate dedicated page.

Database Operations

To accomplish tasks that involve databases, we need to connect with a database and then read and write to it. This can be done in two ways:

i. We can send raw database queries, whose syntax depends upon the type of database we are using (MySQL, Postgresql etc).

ii. We can create a blueprint of database tables in our project, and then use them to read and write to databases.

The latter approach is called model-based approach. These blueprints are called models. Each model corresponds to a table in the database, and each field inside it corresponds to that table column.

For example, a table named Messages, which contains two columns - author and message, of type VARCHAR and max_length of 50 and 100 respectively, will have a model that looks like this:

class Messages(models.Model):

author = models.CharField(max_length=50)
message = models.CharField(max_length=100)

Similarly, we can write any number of models inside models.py file, each corresponding to a table in the database. We do not need to create these tables manually because we already have the blueprints of the tables with us. So, we can simply create tables from our models as well and whatever changes we will do in our models can be updated in those tables.

Do we need to worry about writing the queries to create tables?

No. We just need to define which type of database we are using, and connection credentials in our settings.py file, and Django will take care of the rest, including writing and running queries to create tables. Django has pre-written query structures for each type of database.

CONNECTING TO DATABASE

Example of settings.py for postgresql database:

```
DATABASES = {
  'default': {
     'ENGINE': 'django.db.backends.postgresql_psycopg2',
     'NAME': '<db_name>',
     'USER': '<db_username>',
     'PASSWORD': '<password>',
     'HOST': '<db_hostname_or_ip>',
     'PORT': '<db_port>',
  }
}
```

CREATE AND ALTER TABLES

To create database tables from our models, we need to run these two commands:

i. To prepare migration files containing what needs to be updated on database, we run:

python manage.py makemigrations

ii. To implement above changes, we run:

python manage.py migrate

CREATE AND ALTER DATA IN THE TABLES

After learning the above concept, we can create, alter and delete database tables. Now, we will learn how to feed data into these tables.

Looking back at a model definition, we learn that it is a class. We can make an object from this class. This object will be a table object. This approach is called Object Relational Mapper (ORM) where database table attributes are mapped to object's attributes in Django. Using this object, we can (C)reate, (R)ead, (U)pdate and (D)elete data from these tables. These operations are together called as CRUD operations. Let us learn how to do that in Django one-by-one.

CREATE

To create or save data in table, we simply create an object of table model, assign data to its fields and save them using save() method.

messageObj = Message()
messageObj.author = "Binod"
messageObj.message = "Binod rocks!"
messageObj.save()

We can also initialize these fields while creating the object, like:

messageObj = Message(author = "Binod", message = "Binod rocks!")
messageObj.save()

READ

To read data from the table, we simply use the get() method to pass the required filtering value like following:

messageObj = Message.objects.get(author = "Binod")
print(messageObj.message)

Here, we used the filter author="Binod" to get that specific row of table that has Binod as author. And then we read the message from it.

Caution: Method .get() is used only when we are sure there is just one row that matches our filter. So, in our case, there must be only one row in the Message table that has author named Binod. If there is none or there are more than one, the code will cause an exception and will break.

If we know that there maybe no results or more than one result, we should use filter() instead.

messageQS = Message.objects.filter(author="Binod")

messageQS will be a collection of ORM objects. So, we need to iterate it like:

for messageObj in messageQS:
  print(messageObj.message)

UPDATE

We can update both single object and queryset in Django. To update a single object, we simply assign the object field with a new value and save it, like:

messageObj = Message.objects.get(author="Binod")
messageObj.author = "Manoj"
messageObj.save()

To update a whole queryset, we have update() function:

messageQS = Message.objects.filter(author="Binod")
messageQS.update(author="Manoj")

DELETE

To delete a single object or all the objects in a queryset, we simply use delete() function. For example:

messageObj.delete() # for single object

messageQS.delete() # for all objects in the queryset

What can we do now?

After finishing last section, let's summarize what we have learned so far:

- To connect with database,
- To create and alter database tables,
- To create and alter database tables' data,
- To create and/or locate templates directories to store html files,
- To create and/or locate static files directories,
- To write functions inside views.py files,

- To map functions with url patterns in urls.py files to return html files,
- To link static files content inside html files

Django Template Language

One thing we still need to learn is to write Django template language inside those html files. Remember that we used {% static 'a.jpg' %} in our static files example? This is Django template language.

This language will help us to write conditional statements like if-else to conditionally show data in html pages, to write for loops to show recursive content inside html pages, to load static files data and so on.

In Django template language, we use template tags, like {% %}, {{ }} etc. We will learn only few essential template tags below:

Context Variables

Before this, let's think about something important. When we return an html page, it can only show static content into it - text or static files content like css design, javascript, fonts etc. Now, imagine if we want to show data from our database in the html page. For example, how about showing all messages from author Binod?

This obviously mean that we also need to send this database data alongwith html page, when we return that html page. If we look at that code again which returns the html page:

def myfirstapppage(request):
        return render(request, 'myfirstapp/myfirstapppage.html', {})

we notice a {} in the end. This is called Context-variable. It is a dictionary that is designed to send dynamic content like database data when we return an html page. For example:

```
messageQS = Message.objects.filter(author="Binod")
def myfirstapppage(request):
        return render(request, 'myfirstapp/myfirstapppage.html',
{'messageqs' : messageQS})
```

Now, our messageQS will be available in our myfirstapppage.html template by the name *messageqs*. To print data from this variable, or to use conditional statements around it, or to loop it etc we will use the same template language!

Now, let's go back to learning the template tags which will be much easier now.

{{ }} tag is used to print value of a variable.

{% %} is used for statements like if-else, for loop statements etc.

An example using our *messageqs* variable is:

{% for messageobj in messageqs %}

{{ messageobj.message }}

{% endfor %}
which is similar to writing the following code in views.py:

for messageobj in messageqs:

  print(messageobj.message)

It is just that we use a slightly different language in templates. We can play aound this and create an unordered list in html like:

<ul>

```
{% for messageobj in messageqs %}

<li>{{ messageobj.message }}</li>

{% endfor %}
</ul>
```

An example to use if-else statement:

```
<ul>
{% for messageobj in messageqs %}
{% if message.message %}
<li>{{ messageobj.message }}</li>
{% else %}
<li> No message to show </li>
{% endif %}
{% endfor %}
</ul>
```

PRACTICE 201

programming some small projects

Message App

Let's build a message app where a user can submit messages on one page and see all messages on another page.

Flow

We will build two html pages named sendmessage.html and showmessages.html. The first one will have a form to write message and submit button to save the message in a database. Second page will fetch and show all the submitted messages from the database.

Coding

A. Showing Form Page

i. First, we make a function in views.py of myfirstapp that returns sendmessage.html page to the user where he can type a message and submit it:

views.py

```
def sendmessagepage(request):
    return render(request, 'myfirstapp/sendmessage.html', {})
```

ii. Now, we add a corresponding entry in urls.py for this function:

urls.py

from .views import *

```
urlpatterns = [
url(r'^sendmessagepage', views.sendmessagepage, name='sendmes-
sagepage'),
]
```

iii. We include our myfirstapp urls in main urls.py file of myfirst-project:

```
from django.conf.urls import url
from django.contrib import admin
from django.urls import path, include

urlpatterns = [
url(r'^myfirstapp/', include(('myfirstapp.urls', 'myfirstapp'),
namespace='myfirstapp')),
]
```

iv. We make an html page with a form to submit the message:

sendmessage.html

```
<html>

<head>
<title> Send Message Page </title>
</head>

<body>
<form>
Message: <input type="textarea" name="message">
<input type="submit">
</form>
</body>

</html>
```

v. We run the server using command:

python manage.py runserver

We can now go to 127.0.0.1:8000/myfirstapp/sendmessagepage to see our page.

Note - If you followed all steps correctly and still see Template-DoesNotExist error, it is probably because you forgot to register your myfirstapp in settings.py. It is a common mistake.

Solution:

```
INSTALLED_APPS = [
  'django.contrib.admin',
  'django.contrib.auth',
  'django.contrib.contenttypes',
  'django.contrib.sessions',
  'django.contrib.messages',
  'django.contrib.staticfiles',
  ...
  # Add our app here
  'myfirstapp',
]
```

Message: [   ]   Submit

B. Saving Form Data on Clicking Submit Button

i. When a user submits this form by clicking submit button, form-data will be sent to the server and handled by a function. This function will receive form-data from request and save it in a table named Message, which we will define later in models.py. So, let's write another function inside views.py of myfirstapp to handle this form-data using ORM.

views.py

```
from .models import Message

def sendmessagesave(request):
    # get the message from request
    message = request.POST.get('message')
    #save the message in database using ORM
    messageObj = Message(message=message)
    messageObj.save()
    # Show a success message
    return HttpResponse("Saved!")
```

ii. This function is required to save data in a database inside a table named Message. So, we need to define the model for this Messsage table as well. Let's write this first

models.py

```
class Message(models.Model):
    message = models.CharField(max_length=100)
```

Note: We noticed that we used Message model in views.py that was defined in models.py. This model represents myfirstapp_message table inside our database. Tables are created with the app name as its prefix. So, for our myfirstapp, all tables created via its models.py file will have *myfirstapp_* prefix in them in the database.

iii. Wait! We have not actually created any table yet. We have just coded about it. So, let's create it using:

python manage.py makemigrations

python manage.py migrate

iv. Now, we need to write an entry of our function inside urls.py:

<u>urls.py</u>

from .views import *

urlpatterns = [
url(r'^sendmessagepage', views.sendmessagepage, name='sendmessagepage'),
url(r'^sendmessagesave', views.sendmessagesave, name='sendmessagesave'),
]

v. So, we have created a url that will send the form-data (message) to the function that will save the data inside Message table. But we havent entered this url in form's attribute yet. Let's do it.
Also, forms in Django require an additional thing called CSRF middlewaretoken. They are a security measure to prevent form-submission by an attacker.

<u>sendmessage.html</u>

\<html\>

\<head\>

\<title\> Send Message Page \</title\>

```
</head>

<body>

<form method='post' action='/myfirstapp/sendmessagesave'>

{% csrf_token %}

Message: <input type="textarea" name="message">
<input type="submit">

</form>

</body>

</html>
```

{% csrf_token %} is a template tag that generates csrf token field automatically on page rendering.

C. Showing Saved Data

After saving the messages in database, we need to make a page that shows all the messages.

i. So, let's write the function that will fetch messages from the database using ORM and return these messages alongwith html page in a context-variable (remember, we discussed that context-variables are used to send data alongwith html pages in Django Template Language section?)

views.py

def showmessages(request):

```
messageQS = Message.objects.all()
return render(request, 'myfirstapp/showmessages.html', {'messageqs': messageQS})
```

ii. A urls.py entry to include this function:

urls.py

```
from .views import *

urlpatterns = [
url(r'^sendmessagepage', views.sendmessagepage, name='sendmessagepage'),
url(r'^sendmessagesave', views.sendmessagesave, name='sendmessagesave'),
url(r'^showmessages', views.showmessages, name='showmessages'),
]
```

iii. Create the html page to show messages

showmessages.html

```html
<html>
<head>
  <meta charset="UTF-8">
  <title>Show Messages</title>
</head>
<body>

{% for messageobj in messageqs %}
{{ messageobj.message }}
<br>
{% endfor %}

</body>
</html>
```

That's it. If you now open 127.0.0.1:8000/myfirstapp/showmessages, you will see the messages that you saved before.

www.ingramcontent.com/pod-product-compliance
Lightning Source LLC
Chambersburg PA
CBHW070335240526
45466CB00027B/1991